Mini Milk Maids on the Mooove

By Twins Rianna and Sheridan Chaney

The cover girls are Rianna, Sheridan and Brook-Lodge C Chips. Rianna and Sheridan are the five-year-old twin daughters of Lee and Becky Chaney of Thurmont, Maryland. Chips is a 13-year-old Holstein dairy cow. She is considered a 93-point excellent cow and has produced almost 200,000 pounds of milk in her lifetime. She continues to produce 100 pounds of milk a day; that's about 185 eight-ounce glasses of milk. She is owned by the Thomas and Bonnie Remsberg family of Middletown, Maryland. Chips has been grand champion and supreme champion many times at county and state fairs.

A special Thank You to Miss Betsy (Elizabeth Randall) who let us borrow the John Deere pedal tractor and wagon for our children's book. It belonged to her late son Robbie who passed away at age 7.

To all the *Dairy Farmers* in America,

Thank you for taking such good care of your dairy cows that provide us with delicious and nutritious milk, ice cream, cheese, yogurt, butter and all the great dairy products we enjoy.

Published by Down Under Publications

First Edition, November 2009
©Copyright, Rebecca Chaney, 2009
ISBN 978-0-9818468-1-1

(301) 271-2732
Chaneyswalkabout@aol.com

Edited by Rebecca Long Chaney
Photographs by Kelly Hahn Johnson

Layout and Design by Kathy Moser Stowers

MDA
MID-ATLANTIC
DAIRY ASSOCIATION

A Special Thank You to Mid-Atlantic Dairy Association for helping with this special project.

Oh Pappy, is that really you holding that big brown cow? Did you have to milk cows by hand when you were a little boy? Tell us more about dairy cows and how we use their milk... PLEASE Pappy!

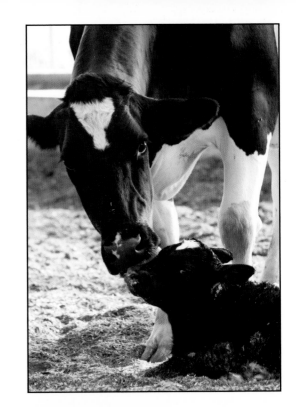

Our Pappy told us that everything on a dairy farm starts in the maternity barn where pregnant cows give birth to their calves. This Holstein calf is just minutes old and already trying to stand on its wobbly legs. She is a heifer, or female, and will grow up on the dairy farm.

We love visiting the Valentine's dairy farm. After a newborn calf gets its first milk, called colostrum, from the mother cow, the calf joins other calves in the nursery barn. The mother cow joins the milking herd to make milk for you and me.

Miss Mandi mixes warm water with powdered milk replacement. It has nutrients and vitamins to keep the calves healthy. It's fun to feed two calves at the same time — it's a race to the finish.

This building is called a greenhouse calf-raising barn. It keeps the young calves warm in the winter and cool in the summer. These newborn calves are wearing calf coats to keep them extra comfy.

We love Brown Swiss calves. It's our Pappy's favorite dairy breed and they came from Switzerland a long, long time ago. Here are the other main dairy breeds and where they came from in the olden days.

Ayrshire
"Scotland"

Holstein
"Netherlands"

Milking Shorthorn
"England"

Guernsey
"Isle of Guernsey"

Jersey
"Isle of Jersey"

Red & White Holstein
"Netherlands"

This grain is awfully heavy... *don't spill it... don't spill it... don't spill it...* Miss Mandi has been taking care of dairy animals since she was a little girl like us. We sure have learned a lot helping her today.

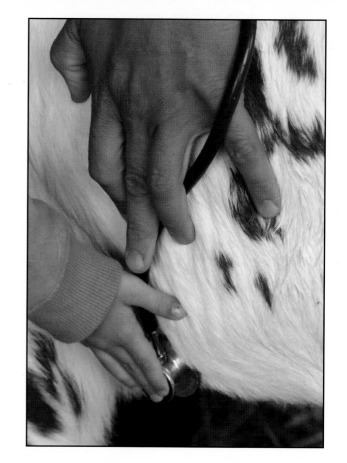

While we were visiting the dairy farm, Dr. Tuck checks on a sick calf. He is a special doctor for animals, called a veterinarian. *Boom, boom... boom, boom... boom, boom...* Dr. Tuck shows us how to listen to the calf's heartbeat with a stethoscope.

In the milking parlor Miss Mandi shows us how they milk cows with machines. It's awesome watching the milk squirt out. Did you know the average cow drinks about a bathtub of water a day to produce 90 glasses of milk? This sure looks easier than milking by hand!

Mr. Richard is a professional truck driver and picks the milk up from Vales-Pride Farm every other day. He takes a sample of the milk from the milk tank for testing at the processing plant to ensure its quality.

The Pee Wee Pretty Cow Contest at the county fair is always fun. Our theme was "Eat Dairy Products and Dance." Samantha, the Jersey calf, is so cute. The calcium in dairy products builds strong bones that make us better dancers.

"This is really not my color; I would prefer a darker red."

Our friend Joseph is "Moocheal Phelps." He and his Ayrshire calf, Brooke, just won eight gold medals for the United States. The winners were all dairy products including milk, butter, ice cream, cheese, yogurt, sour cream, cottage cheese, and whipped cream.

At one of the first Lewistown Mt. View 4-H Cloverbuds' meetings, we made butter in a jar and ice cream in a bag. 4-H is the largest youth group in America for youth ages 8 to 18. Cloverbuds are for children 5 to 7 and teaches kids, like us, about all the neat opportunities in 4-H.

Lucia thinks the butter is finger-licking good. Marshall is convinced that making butter takes way too long. Cadin and Joseph are excited because with a few more shakes of their jar, they will have creamy butter to eat.

Making ice cream was no easy job. We squished the bags of ingredients and ice for almost 10 minutes. Our hands got so cold, but it sure tasted delicious. Brady liked his too, until our dog, Jaxson, decided he liked it more.

The 4-H Cloverbuds took a June Dairy Month tour of South Mountain Creamery. There, we learned what happens to milk once it leaves the farm. For the past 70 years dairy farmers have celebrated June Dairy Month. We helped Mr. Seth package ice cream.

South Mountain Creamery

Black Raspberry

Ice Cream

Milk, Cream, Sugar, Nonfat milk powder, egg, Pure & Artificial Vanilla, Natural Stabilizers. *** MAY CONTAIN NUTS OR NUT OILS

Fresh from Our Farm to Your Home

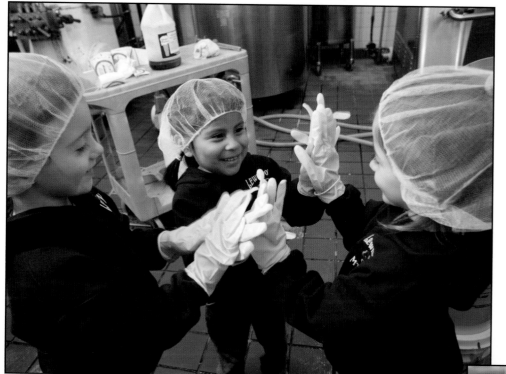

At South Mountain Creamery they turn the milk into 35 flavors of ice cream. Things in a creamery must be very clean, so we wear hair nets and plastic gloves. Sophia pours the ingredients into the ice cream maker for the next batch. Mmmm... we make really good ice cream.

After we made ice cream, we played with Sophia on her grandparents' dairy farm. The farm supplies all the milk for the creamery. They milk 220 cows two times a day and get lots of milk for their customers.

At South Mountain Creamery 4,000 half-gallon milk bottles are filled every Monday, even on holidays! Cadin and Joseph put bottles on the wash line. They are cleaned in a huge bottle washer that heats up to 151 degrees Fahrenheit — wow, that's hot! The hot water sanitizes to kill any germs in the bottles!

Once the clean bottles are filled with milk that has been pasteurized, Marshall, Wade, Brady and Emma carefully place the bottles of milk into the crates. The milk is ready to be delivered door-to-door to 4,000 homes in four states.

Our work is done! Now, it's time for our favorite treat — ice cream! Make sure you drink your milk and eat your dairy products to have strong bones and healthy teeth!

Not my ice cream!!! The farm dog, Vinnie, surprises me when he takes a big bite of my ice cream. I hope Vinnie likes my cotton candy flavor. I guess dogs need their calcium, too.

Pappy, this feels funny; it's so soft. We sure did learn a lot about dairy cows and milk this year. Pappy, bet you're happy you don't have to milk by hand like you did as a little boy; Rianna, my hand hurts. Are you ready to have a try?

Fun Dairy Activities

MILK MUNCHIES

Find the letters in the milk cartons to complete the DAIRY DELICIOUS, good to eat treats!

G O T M I L K ?

$\overline{3}$ $\overline{19}$ G $\overline{12}$ $\overline{13}$ $\overline{15}$

$\overline{1}$ $\overline{19}$ O $\overline{4}$ $\overline{16}$ $\overline{18}$ $\overline{6}$

$\overline{7}$ $\overline{18}$ $\overline{11}$ $\overline{17}$ $\overline{12}$ T $\overline{9}$ $\overline{12}$ $\overline{15}$ $\overline{15}$ $\overline{18}$ $\overline{13}$

M $\overline{12}$ $\overline{10}$ $\overline{10}$ $\overline{16}$ $\overline{17}$

$\overline{7}$ $\overline{12}$ $\overline{5}$ $\overline{5}$ I $\overline{17}$ $\overline{14}$

$\overline{2}$ $\overline{11}$ L $\overline{15}$

$\overline{6}$ $\overline{8}$ $\overline{11}$ K $\overline{18}$

?

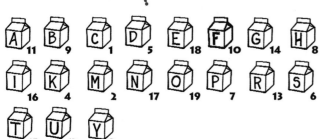

A 11 B 9 C 1 D 5 E 18 F 10 G 14 H 8

I 16 K 4 M 2 N 17 O 19 P 7 R 13 S 6

T 15 U 12 Y 3

Provided by Mid-Atlantic Dairy Association; Developed by Midwest Dairy Association.

Cheeses of the World

Notice where these cheeses come from. Today, they all can be made in the United States. Find the cheeses:

CAMEMBERT *(France)*
GORGONZOLA *(Italy)*
BLUE *(France)*
BRIE *(France)*
BRICK *(America)*
FETA *(Greece)*
CHEDDAR *(Great Britain)*

MOZZARELLA *(Italy)*
COLBY *(America)*
LONGHORN *(America)*
PROVOLONE *(Italy)*
ROMANO *(Italy)*
SWISS *(Switzerland)*
PARMESAN *(Italy)*

GERMANY
FRANCE
GREAT BRITAIN
AMERICA
ITALY
GREECE
SWITZERLAND

X	T	R	E	B	M	E	M	A	C	R	P	D	H
T	Q	B	Y	A	B	Q	J	H	K	Z	X	M	S
S	K	C	I	R	B	F	E	T	A	O	C	Y	A
C	R	I	K	O	B	D	E	I	G	L	O	L	E
W	L	C	F	M	D	L	F	U	N	X	L	J	N
N	O	T	H	A	F	P	W	Z	L	E	B	E	O
A	N	G	R	N	G	J	E	I	R	B	Y	K	L
P	G	Z	D	O	E	A	C	A	G	C	H	S	O
T	H	U	N	V	M	O	Z	B	M	F	S	N	V
G	O	R	G	O	N	Z	O	L	A	I	Y	I	O
D	R	V	Q	E	O	K	X	J	W	D	L	Z	R
U	N	L	A	M	B	N	A	S	E	M	R	A	P

Glossary

Calf Coat - A cotton coat specifically made to fit a newborn calf to keep its temperature regulated and warm in cool or cold weather.

Colostrum - The first milk that a mother cow produces and the most nutritious food for a newborn calf to drink to help the calf build its immunities against sickness or disease.

Dairy Farmer - A person who raises dairy cattle for the primary purpose of milk production.

June Dairy Month - Celebrated since 1939, June Dairy Month recognizes dairy farm families for their contributions to the health and economy of the United States.

Milking Parlor - The structure on a farm where cows are milked with milking machines.

Pasteurized - A process that kills any possible germs in milk.

Stethoscope - The instrument a veterinarian or animal doctor uses to listen to an animal's heartbeat.

Recipes

Butter in a Jar

You will need
A small baby food jar
A pinch of salt
1 teaspoon heavy whipping cream
Crackers

1) Place a teaspoon of heavy whipping cream in the jar with a pinch of salt.
2) Shake jar until butter is formed.
3) Spread creamy butter on cracker and enjoy!

Ice Cream in a Bag

You will need
Ice cubes
1/2 cup kosher salt
1/2 teaspoon vanilla extract
1 cup half-and-half milk

2 tablespoons sugar
1 resealable sandwich bag
1 gallon-sized resealable bag

1) Fill the large bag with ice. Add the salt.
2) Mix half-and-half, sugar, and vanilla in the small sandwich bag. Seal tightly.
3) Place the small bag inside the large bag, seal tightly.
4) Wrap a dish towel around the big bag and squish until the ice cream thickens. (About five to ten minutes)
5) Eat the ice cream out of the bag or place in a bowl. Yum, Yum!!

Fun Dairy Facts

In one day, an average dairy cow produces enough milk to make eight gallons of milk, or seven pounds of cheese or two pounds of butter --- The largest milk producing states in the U.S. are California and Wisconsin --- The largest ice cream producing states include Pennsylvania, California, Indiana, Texas, Illinois and Minnesota --- A cow burps her food back up and chews it for about eight hours every day; this is called chewing her cud --- The first cow arrived in America in 1611 --- There are nine million dairy cows in the U.S. today --- A cow has four stomachs and 32 teeth --- The average person eats nearly 50 slices of pizza with cheese per year --- Milk is also used to make glue, paint and plastics.

For Kids and Teachers

www.dairyspot.com – Dairyspot is the official Web site of Mid-Atlantic Dairy Association. Mid-Atlantic is one of 18 state and regional dairy product promotion organizations working under the umbrella of United Dairy Industry Association on behalf of dairy farmers. They provide education resources and other information for children and adults.

www.4-h.org – 4-H has grown into a community of 6 million young people across America learning leadership, citizenship and life skills. 4-H can be found in every county in every state, as well as the District of Columbia; all U.S. territories and over 98 countries around the world. 4-H'ers participate in fun, hands-on learning activities, supported by the latest research of land-grant universities.

www.nutritionexplorations.org – Get tips on fun ways to cook with kids, advice from nutrition experts, art contests and recommended reading.

www.dairyfarmingtoday.org – Educates the public about how today's dairy farmers care for their animals and their land, while growing healthy farming businesses for future generations. The site tells about dairy producers' hard work every day to provide safe, wholesome and nutritious milk.

It also focuses on the many positive contributions the dairy industry makes to rural America. A virtual dairy farm tour is available on the site.

www.ilovecheese.com – Everything you want to know about cheese, including recipes, entertaining with cheese, shopping for cheese, and more.

www.usdairy.com – The Innovation Center for U.S. Dairy's Web site explains the dairy industry's road map to greenhouse gas reduction and related projects. This site can serve as a primary resource for information about dairy sustainability.

www.moomilk.com – Take a virtual tour of milk from the cow to your table, play fun games, take the Moo Milk quiz, get tasty recipes and much more.

www.nationaldairycouncil.org – Nutrition and product information, health professional resources, tools for schools and hands-on activities, recipes, health tips and more.

www.southmountaincreamery.com – Offers fresh, all-natural dairy products. Has product and home delivery information along with recipes and online ordering. The Sowers and Brusco families encourage free daily farm experiences.

www.maefonline.com – The Maryland Agricultural Education Foundation focuses on agricultural education. An elementary "Ag in the Classroom" workshop is held every summer and provides teachers with hands-on lessons matched to the State Curriculum and resources ready to be taken into the classroom. (www.agclassroom.org)

www.gotmilk.com – Official "got milk?®" Web site. Filled with recipes, articles, crafts and interactive games related to the health benefits of milk.

www.hilmarcheese.com/CowTour.cms – From California's Hilmar Cheese comes this virtual tour of cheese production. Get an up-close view of every step of the process including manufacturing, recycling, packaging and sales.

www.kellyhahnphotography.com - Kelly Hahn Johnson is an award-winning photojournalist with more than 20 years of experience. See more photos, videos and testimonials here.

www.discoverdairy.com – A lesson series that teaches upper elementary students where dairy foods come from. Lessons include animal health, milk safety, environment, community and nutrition. Videos, games and materials are available as free downloads.

www.rebeccalongchaney.com – View photos and video footage of Rebecca, Rianna and Sheridan's mission to teach young people about agriculture. Order more books here.

Touting themselves as "Agricultural Warriors," Rebecca, Rianna and Sheridan are proud they are taking their agricultural message on the road. In 2009, the trio visited eight schools reaching 1,300 elementary students. Their interactive 45-minute ag presentation has Thurmont Primary School students dubbing them the "Cow Family." In addition to promoting farm life and agricultural products, Rebecca is a freelance journalist and inspirational speaker. The Chaneys are back in Maryland living on the old Long View Farm, the family farm, that had been in Rebecca's family 100 years. Owned today by Elizabeth Randall, Randall Land & Cattle Company has given Rianna and Sheridan the opportunity to be the fourth generation involved in agriculture on the same farm. Lee Chaney oversees 80 registered Hereford beef cows. The girls look forward to their first Brown Swiss dairy calves due to be born in spring, 2010.

The twins enjoy all-day kindergarten, love church activities, play softball, ride their bikes whenever they can, and just started horseback riding lessons. They especially love to help their daddy check the cows, watch newborn calves and play in the hay with cousin, David. To find out more information about their ag presentations or Rebecca's speaking availability, contact her at (301) 271-2732 or email chaneyswalkabout@aol.com, info@rebeccalongchaney.com or check out her Web site at www. Rebeccalongchaney.com.

Make sure you collect all the books in the Chaney Twins' Series, Here's #1!!

Kathy Moser Stowers lives with her husband and children on her family's second farm near Jefferson, Maryland. She has 24 years of experience in design and layout at The Frederick News-Post. She has shared her talents and expertise with local, state, national and international groups, organizations and events. She specializes in brochures, flyers and business cards.

For help in designing brochures, pamphlets, books, business cards, etc., contact Kathy at Wakstowers@aol.com or call (301) 748-9112 or (301) 371-9306.

Award-winning photographer Kelly Hahn Johnson not only is known for her unique photojournalism style but also her approach to portraiture. She's won countless awards and her images have been featured in local, state and national publications. She loves observing people, moments and emotions, creating photos that will be treasured for years to come, like the priceless images she's captured of the Chaney twins over the years. She lives with her husband Blane and son Brady in a century-old renovated house in Sharpsburg, Maryland.

Visit her online gallery at kellyhahnphotography.com. Contact Kelly at info@kellyhahnphotography.com or call her at (240) 285-3677.

A Special Thanks to all the people and their animals who made this book possible!

The Valentines of Vales-Pride Farm • The Iagers of Bull Dog Holsteins
The Thomas & Bonnie Remsberg Family of Brook-Lodge Farm • South Mountain Creamery
The 4-H Cloverbuds including Wade Stowers, Marshall Hahn, Emma Ford, Cadin Valentine, Joseph Hubbard,
Lucia Runkles, Brady Johnson, Sophia Brusco • The Keilholtz Family of Glad-Ray Farm
The Mayers of Stoney Point Farm • Keilholtz Trucking • The Smiths of O-C-S Dairy
The Hahns of Wind Swept Farm • The Guytons of Guy-Dell Farm • Dr. Paul Tuck
Cindy Warner of Round Hill Acres and an extra special thanks to Pappy and Memaw Long,
LaLa Chaney, and Mommy and Daddy who teach us about farm animals.